Chinese Folklore Picture Book Series

The Secret of the Long-haired Girl

Written by Liu Xushuang
Illustrated by Wang Xiaoxiao
Translated by Phyllis Ang

At the beginning of this story, the mountain monster was sleeping in the belly of the mountain.

Tens of thousands of years have passed while he was asleep.

The long-haired girl was climbing up the cliff with a bamboo basket on her back.

She wrapped her beautiful black braids around her neck and reached for a clump of flowers on the cliff.

When she uprooted the flowers, spring water gushed out of the rock wall.

There was a low roar from behind the rock wall, which frigthened the birds into flight.

Beasts fled. Bushes and leaves shook. And spring water splashed.

The long-haired girl suddenly felt very light.
A big hand had picked her up.

It was the mountain monster!

The mountain monster looked her in the eyes and shouted, "Se-cre-t!"

The long-haired girl stared back in his crystal-clear blue eyes, and saw the secret of the mountain monster: a wonderful dream world, guarded by the mountain monster with spring water.

At night, the girl couldn't fall asleep when recalling that clear spring.

Animals and crops were dying for thirst, and so were people. This dry land needed water desperately.

A gust of wind came from the other side of the mountain, and the wind carried the roar of the mountain monster.

"It's a secret... a secret... I will take your most precious thing if you tell anyone else!"

The long-haired girl buried her head deep in the quilt.

The next day, the girl walked towards the mountain.

"But what about the secret of the mountain monster?

What would he take away from me? My life?"

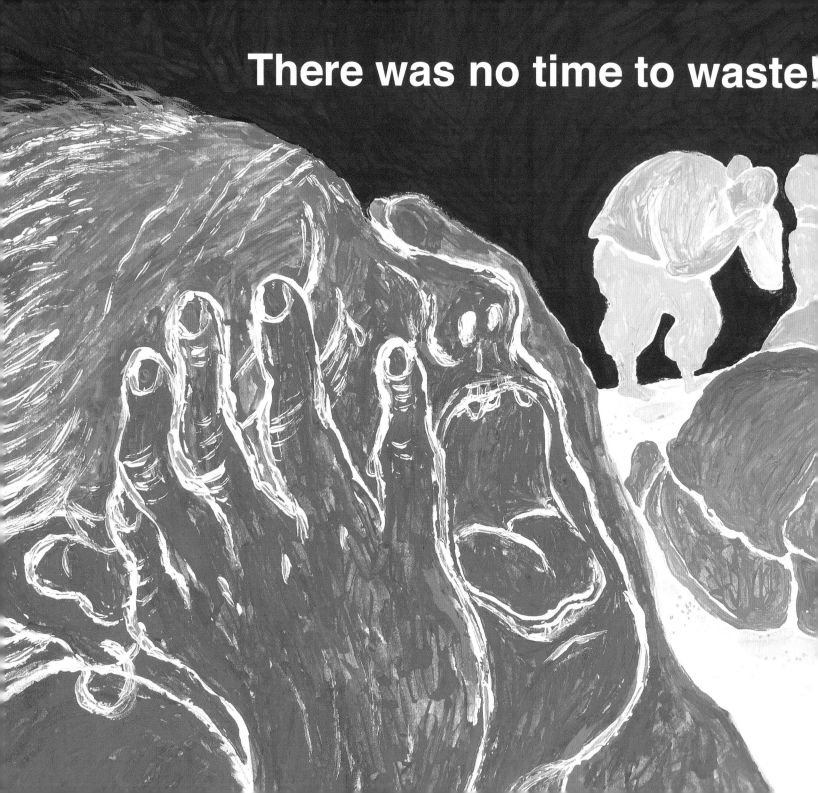

The long-haired girl jumped up and ran towards the mountain.

As she ran, she shouted at the top of her lungs,

" There's water in the mountain! Spring water! "

She climbed up the cliff again and reached out to the clump of flowers.

The mountain monster looked at her silently.

The spring water gushed out again, and the spring became bigger and bigger, turning into a waterfall.

The long-haired girl fell into the water and sank.

The spring water was so cool, so gentle ...

She felt as if she had fallen into the mountain monster's dream world, sinking deeper and deeper.

Her life would probably disappear like this.

Then, a pair of hands lifted her up.

She knew,
these were the mountain monster's hands.

These hands were getting smaller,
and weaker.

The spring water surged with delight, and cheers came out from the foot of the mountain.

The long-haired girl suddenly realized that she was still alive, but she would never see the mountain monster again.

"Mountain monster, mountain monster..." she called softly.

A gust of wind blew, as if in answer.

The wind lifted her long, long hair, breaking it off inch by inch to fly away in the wind.

Was it the mountain monster who took her long hair?

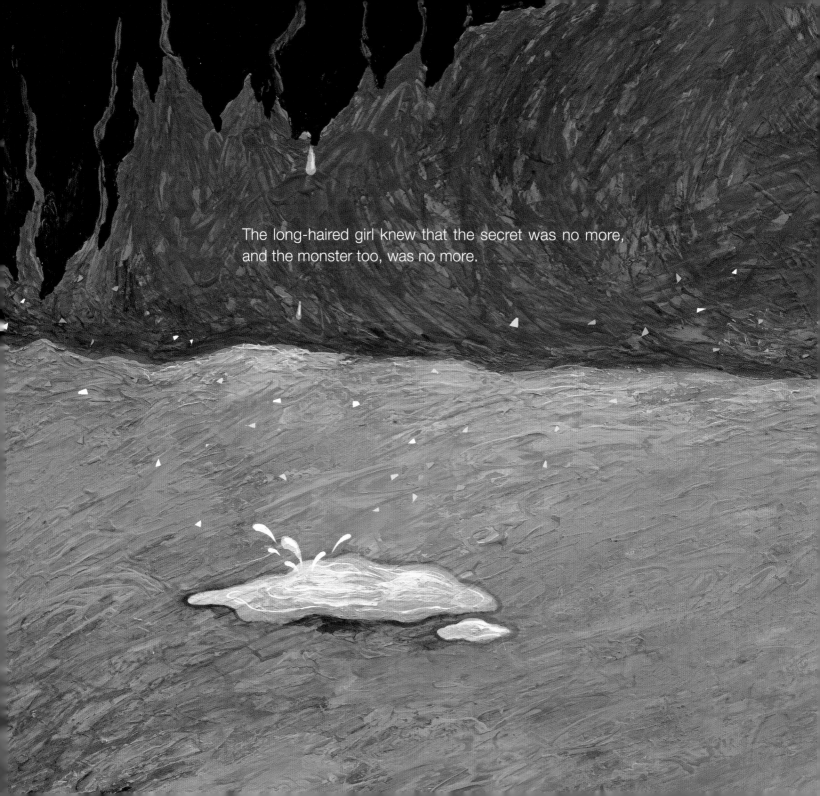

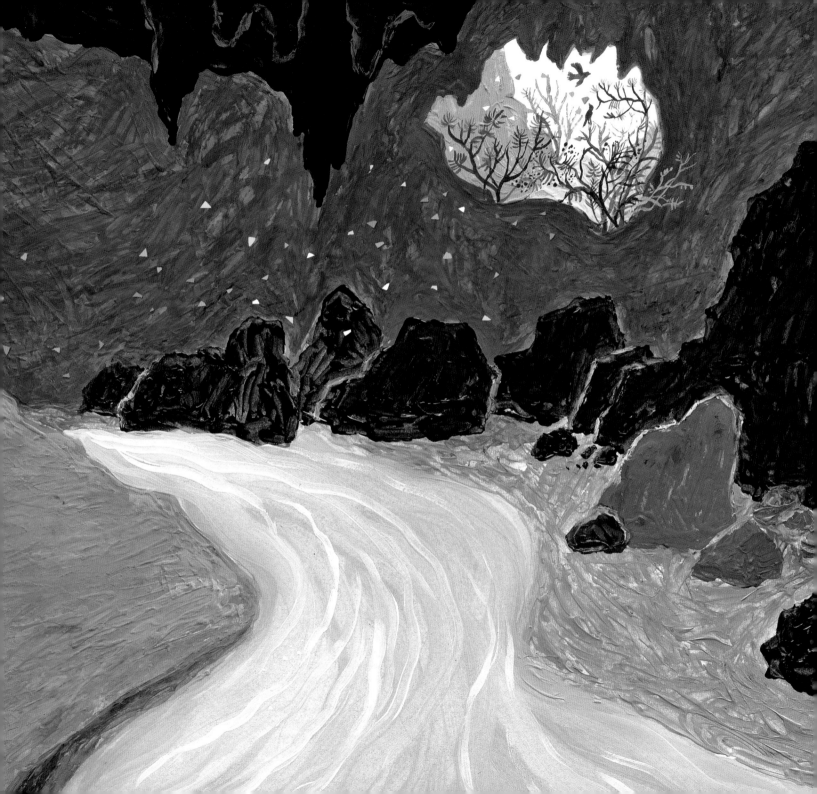

Chinese Folklore Picture Book Series

The Secret of the Long-haired Girl

Written by Liu Xushuang
Illustrated by Wang Xiaoxiao
Translated by Phyllis Ang

Editor Jane Yang
Graphic Designer Lily Tang
Print Production Eric Lau

Chung Hwa Book Co., (Singapore) Ltd., 2021

Published by Chung Hwa Book Co., (Singapore) Ltd.
211 Henderson Road, Singapore 159552

http://www.chunghwabook.com.hk

First Published in December 2021

Printed by Elegance Printing & Book Binding Co., Ltd.

Distributed by SUP Publishing Logistics (H.K.) Ltd.

ISBN 978-981-18-2230-8

Text copyright © Liu Xushuang
Illustration copyright © Wang Xiaoxiao
Originally published in Chinese by Jieli Publishing House Co., Ltd. in collaboration with Shanghai Yishao Creative Studio
English edition translation © Chung Hwa Book Co., (Singapore) Ltd. 2021
Published by arrangement with Jieli Publishing House Co., Ltd. P.R.C.
China-ASEAN Children's Book Cooperation Project of Jieli-ASEAN Children's Book Alliance
ALL RIGHTS RESERVED